SOCIETY OF PETROLEUM EVALUATION ENGINEERS

SOCIETY OF PETROLEUM EVALUATION ENGINEERS

DISCUSSION AND GUIDANCE ON ETHICS

JANUARY 2018

SOCIETY OF PETROLEUM EVALUATION ENGINEERS ETHICS COMMITTEE

Contents

Letter from the Committee

January 23, 2018

It has now been over ten years since the publication of *Discussion and Guidance on Ethics Prepared for the Society of Petroleum Evaluation Engineers,* by the Society of Petroleum Evaluation Engineers Ethics Committee (Committee). The intervening years have seen three U.S. Presidents, four British Prime Ministers, explosive growth in unconventional oil and gas resources, a major offshore well blowout, and a near meltdown of world financial markets. The Society of Petroleum Evaluation Engineers has expanded its membership well beyond North America and is at present growing rapidly in Europe.

Although ethical principles themselves have not experienced this level of change, the context in which they are practiced has changed. The SPEE Ethics Committee believes that this is an opportune time to revisit and refresh the Society's ethical guidance. Given the speed with which society and technology is changing, the Committee has taken the decision to review this document annually and issue updates as warranted.

It is our hope that you will find this document a useful tool both in avoiding ethical challenges, and in addressing the ethical dimension of your day to day practice of petroleum reservoir evaluation. We encourage you to review this document annually and provide the Committee with your feedback for use in our annual review.

Sincerely,

SPEE Ethics Committee

Tim Gilblom, P.E.
Charles Gleeson, P.E.
Michael Horne P.E
Jay Thrash, P.E.

Chair:
Tom Collier, P.E.

Introduction

Professional ethics in a free society can only be achieved when individual professionals conduct themselves with honesty, accountability and goodwill towards society at-large. Ethical application of technical work is at the heart of the practice of evaluating oil and gas properties.

An ethical evaluator displays the trait of honesty by being committed to the transparent explanation and exposure of facts, assumptions, results and conclusions and is not dependent upon the end use of the evaluation nor upon the situation or the culture in which it will be used. An ethical evaluator displays the trait of accountability by being committed to a reconciliation between actual and estimated outcomes even when such a reconciliation will result in a loss in professional standing and/or career development. An ethical evaluator displays the trait of goodwill towards society at-large when he or she seeks to bring about positive and good financial and industrial activity (sometimes by condemning the commerciality of the contemplated activity).

This document has been prepared by the SPEE in order to provide non-prescriptive guidance on the ethical aspects of the professional practice of petroleum evaluation. The SPEE specifically disclaims responsibility for any other use of this document. Further, nothing in this document should be considered to be an attempt to provide legal advice, even in areas where ethical performance and the law overlap. The SPEE strongly recommends that all legal advice regarding ethical issues should be secured from appropriately certified legal professionals.

Principles of Acceptable Evaluation Engineering Practice[1]

Article I Fundamental Canons

Engineers, in the fulfillment of their professional duties, shall:

- Hold paramount the safety, health and welfare of the public.
- Perform services only in areas of their competences.
- Issue public statements in an objective and truthful manner.
- Act for each employer or client as faithful consultants or trustees.
- Avoid deceptive acts.
- Conduct themselves honorably, responsibly, ethically and lawfully so as to enhance the honor, reputation and usefulness of the profession.

Article II General Practice

1. The Society may adopt and issue from time to time guidelines for the preparation of reserve estimates and evaluation reports.
2. Members will ascertain that their reports include a full disclosure of the reserve definitions used, either by reference to a known industry set of definitions or by actual inclusion of the definitions used. The reference or inclusion of the definitions shall be clearly presented so as to enable the reader to easily ascertain the definitions used.
3. If the member preparing the report has a vested interest in the properties being evaluated, the nature of this interest shall be disclosed.

---

[1] Source: https://secure.spee.org/joining-spee – "Important SPEE Documents".

Article III Relation of Members to the Public

1. Members will make oral and written statements that are honest and fair, avoiding exaggeration and sensationalism.
2. Members will control the use of maps and reports to assure they are used only for legitimate purposes. For example, limitations in the data relied upon or the method of analysis employed or the assumptions made must be explained. Also the specific use of the maps and reports will be defined.
3. Members will give professional opinions, prepare reports or give legal testimony only after adequate preparation. The extent of that preparation shall be disclosed.
4. Members will publish business and professional announcements, but shall not advertise their work or accomplishments in a self-laudatory or conspicuous manner. Members should not misrepresent their experience nor professional or academic qualifications.
5. Members shall report violations of this Code of Ethics to the appropriate professional bodies.

Article IV Relation of Members to Employer and Client

1. A member shall protect, to the fullest extent possible, the interest of his employer or client so far as it is consistent with the laws of the state, the public welfare, and professional obligations and ethics.
2. A Member will not use or divulge, directly or indirectly, any client's or employer's confidential information without express written consent. To do so is unethical and may constitute a theft.
3. A Member retained by one client will not accept, without that client's consent, an engagement by another if the interest of the two is in any manner conflicting.
4. A Member who has made an investigation for any employer or client will not seek to profit economically from the information gained unless permission to do so is granted or until it is clear that there can no longer be a conflict of interest with the original employer or client.
5. A Member shall not seek or accept a contingent fee arrangement for preparing a report, giving a professional opinion or providing legal testimony. This destroys the credibility of the product.
6. Member shall not accept a concealed fee for referring a client or employer to a specialist or for recommending petroleum evaluation services other than his own. Such an undisclosed payment is a kick-back.

Article V Relation of Members to Peers

1. A Member will not falsely nor maliciously attempt to injure the reputation or business of another person.
2. A Member will freely give credit for work done by others, will refrain from plagiarism in oral or written communications and will not knowingly accept credit rightfully due another person.
3. A Member will endeavor to cooperate with others in his profession and will encourage the ethical dissemination of petroleum evaluation knowledge.

Article VI Duty to the Society

1. In making application to become a member or continuing as a member in the Society, a member agrees to uphold the NSPE Code of Ethics for Engineers and these Principles of Acceptable Evaluation Engineering Practice by precept and example.
2. A Member of the Society will aid in preventing the election of a person to membership in the Society who does not abide by the NSPE Code of Ethics for Engineers and these Principles of Acceptable Evaluation Engineering Practice or who does not have the required education and experience.

Ethical Considerations for Petroleum Evaluators

Introduction

The petroleum evaluation engineer (Evaluator) is subject to standards of performance and standards of behavior that recognize the simultaneous duties to the public, to the client or employer, and to fellow Evaluators. In many ways these duties are similar to other professional disciplines. Uncertainties inherent in petroleum evaluation lead to a wide range of possible outcomes. Given that actual results may vary considerably from early estimates, the integrity and competence of the estimator is critical. Our ethics must be above reproach. They must be internalized and lived out both in our practice of petroleum evaluation (standards of performance) as well as in our personal and professional lives (standards of behavior). This document will provide guidance in seven areas of evaluation ethics.

- Relation to Society
- Standards of Performance
- Competency
- Conflict of Interest, Independence and Disclosure
- Professional Work Product
- The Environment
- Duty to Disclose

Petroleum evaluators may be either employees of their companies or independent consultants. The ethical guidance provided in this document generally applies to both, but is not identical to both groups. Where ethical guidance applies to both groups, the term evaluator is used. Where the ethical guidance is specific to one group or another the terms employee/employer or consultant/client are used for employees and consultants, respectively.

Differences in who controls the work results in different ethical responsibilities between employees and consultants. The consultant, as an independent contractor, is in full control of the work and is therefore ethically required to deliver a work product that meets the need for the use intended by the client. The consultant is responsible for the content and the accuracy of the work product, and carries an obligation to obtain required information from the client. The consultant must take steps to assure that client-provided data is reasonable. Under certain circumstances the consultant may be ethically required to disclose concerns surrounding the data provided to higher level client stakeholders. Where the consultant is unable to obtain, or reasonably suspects that the client information provided is false or inaccurate, the consultant may be required to decline the engagement.

Since the company is responsible and liable for the work of its employees, an employee is responsible for faithfully following the company's procedures and standards of performance. Should a question or concern arise in the course of employment, the employee should direct that concern to the appropriate level of management of the company. When employment activities are related to public disclosures (e.g., reserves disclosures) a higher standard attaches. Since employees should still hold paramount the public welfare, an employee who reasonably determines that the employer is intentionally releasing false or erroneous information is required to report such activities to higher level management or the company's board of directors. Continuing ethical violations by the company may require an employee to resign from the company's employment.

The Petroleum Evaluator's Relation to Society

For one to understand the responsibility that an individual practicing as a petroleum evaluation engineer has to our community and to the larger collective, one can look to the SPEE Mission Statement.

> "Recognizing that Petroleum Evaluation Engineering is a specialized field, the Society is dedicated to the promotion of professional growth of the membership and to the advancement of the profession of Petroleum Evaluation Engineering by demonstrating by example the highest standard of ethics, by promoting continuing education of our membership and by education of the public in the area of oil and gas reserve definitions, reserve evaluations, and fair market value."[2]

An engineer pursues their discipline with a code of ethical behavior and principles of acceptable evaluation engineering practice. Among the fundamental canons are that an engineer shall hold paramount the safety, health and welfare of the public and shall perform services only in the areas of their expertise. This is detailed in the SPEE Code of Ethics for Engineers and the Principles of Acceptable Evaluation Engineering Practice can be found in the Appendix. The National Society of Professional Engineers has established a Code of Ethics that is similar in nature.

Ethics can be described as philosophical concepts and principles that guide an individual's behavior and judgment to identify and react in a morally correct manner in the course of everyday life. Since the beginning of civilization whereby groups of people or communities were established, rules or codes of conduct were in place setting out the greater good in some manner for that society. Throughout history, whether the Greeks, other societies or religious beliefs laid a foundation for understanding right from wrong and moral living. Adam Smith is

---

[2] Society of Petroleum Evaluation Engineers: https://secure.spee.org/about-us: Accessed May 9, 2017.

considered the father of modern economics and a proponent of free market capitalism and wrote *The Wealth of Nations*, but as a member of the Scottish Enlightenment period he spent much of his life observing and writing about his fellow man reacting to each other. Smith as a moral philosopher wrote The Theory of Moral Sentiments in which he outlines the science of man as a social being and how individuals function and relate to each other in a community. He believed an individual's moral philosophy and nature are a combination of empathy and emotional connection to one another that is reflected in one's judgment and reaction to a person's behavior in a particular setting. Smith believed individuals make decisions and react primarily with self-interest or preservation in mind. In essence people judge their own actions and other's actions by their personalized belief system which can and does include the general beliefs of the community at large. One may reason this is why government law and regulation occur, and organizational standards of conduct and procedure arise from societies.

Engineering applies science and mathematics to our physical world to provide useful products to mankind. This is why an ethical professional engineer's first duty is to place the safety of the public above all else. However, in assessing the ethical behavior as it relates to petroleum engineering sub-disciplines such as petroleum reservoir engineering or to the SPEE members' sub-discipline of petroleum evaluation engineering, one needs to understand the range or scope of work involved and the purpose of the resultant product. For instance, the evaluation engineer may need to state an opinion about the value and condition of reserve estimates for the United States Securities and Exchange Commission (SEC) in connection with a public company's assets. An engineer practitioner as a member of the SPEE would accept the assignment knowing the resultant product must and will be done in a competent and ethical manner with proper credentials and licenses using best practices as guidance in applying the requirements of applicable law and regulations particularly associated with the SEC. If the assignment is to determine fair market value of reserve estimates for local or state taxation for a specific property, a practitioner would accept the assignment knowing the resultant product must and will be done in a competent and ethical manner with proper credentials and licenses using best practices as guidance in applying the fair market value concept in accordance with the laws and regulations of the proper jurisdiction.

The advent and explosive production increase of unconventional resources and the remarkable changes in technology that have occurred in petroleum engineering in the past two decades have provided the practitioner of this discipline with new challenges in estimating reserves, recovery rates and ultimately their value. This is why, along with experience, continued education in technology and staying abreast of rules and regulations pertaining to the evaluation engineering practice is very important and a central theme in the SPEE mission. SPEE provides opportunities for its members and peers to interact in forums, seminars and conventions along with publications pertaining to recommended evaluation practices, current technology, and common definitions to ensure consistent communication and a quality work product. In order to make the public aware of sound evaluation engineering practice, the SPEE provides resources and membership information as a tool for the general public and the community involved in the oil and gas industry. Likewise, the petroleum evaluator should react in an ethical manner to protect the investor, stockholder, and those involved in financing or other business concerns from deception or misrepresentation even if this position is contrary to a client's interest. Intentional perversion of the truth has no place in the professional practice of engineering for the client, the company employer, or the community in which one does business.

Petroleum Evaluation Standards of Performance

This section is not intended to provide technical guidance for performing petroleum evaluations but rather to offer an ethical framework to be applied to technical efforts.

Evaluators must:

- Meet the standards of expertise in petroleum evaluation as described in the Competency section of this document, or have their work reviewed by an expert.
- Know the relevant resources and procedures to make a proper evaluation. These may be SPEE monographs, SEC guidelines, SPE documents, COGEH (for Canada), recommended practices applicable to particular jurisdictions, standard industry practices, or any source that the expert deems relevant to the task.
- Employ those resources and procedures in the most reasonable manner.
- Emphasize thoroughness and rigor over expediency.
- Use best professional and technical judgment in weighing the results of an evaluation.
- Be prepared to fully explain and defend the full process used in an evaluation, including the areas of uncertainty.

Competency

Competency (or competence) is defined as the ability of an individual to do a specific job task properly. In practical characterization of an individual's ability, it is generally recognized that there is no clear dividing line between incompetency and competency. Rather, there is a gradual progression from beginner to expert. According to the Business Dictionary[3], an expert is a professional who has acquired knowledge and skills through study and practice over the years, in a particular field or subject. The key words in that definition are "through study and practice" and "in a particular field or subject".

With regard to SPEE's area of practice, there must be a way to assure that the task of petroleum evaluation is done correctly by evaluators at all levels of competency as they progress toward

3 http://www.businessdictionary.com/definition/expert.html

expert status. New practitioners must seek out guidance from experts; experts must make themselves available to guide new practitioners.

We must recognize that we are not expected to be experts in everything, even within the relatively narrow subject of petroleum evaluation. Competence is task specific. We should clearly understand what our individual areas of competence are, and what they are not. While we may use cues from others to assess our level of expertise, it is ultimately our own responsibility to know our level of competency and to act accordingly.

An expert or someone who is fully competent:

- Understands the technical theory behind a task
- Knows the methods to apply the theory
- Can choose the best method(s) to employ for the specific task
- Can assess the "reasonableness" of the results of the evaluation
- Understands the context and potential impact of those results
- Can clearly explain all of the above to a "less expert" individual
- Can mentor a less experienced individuals to improve their competence

We can weigh our abilities against these guidelines to assess our level of expertise. We can also ask ourselves some practical questions, such as:

- Would I be comfortable defending my work in a contested situation in which I represent either a buyer or seller of a property?
- Would I expect generally accepted experts to consider me to be a peer?
- Have I ever been sought out by experts to provide my technical opinion?
- Have I mentored less experienced individuals to help them improve their competency?

Conflict of Interest, Independence & Disclosure

When we talk about a conflict of interest we are not discussing the conditions in which conflicting interests exist, rather we are discussing cases where the interest of the evaluator, if pursued might prevent the faithful fulfilment of their professional responsibilities.[4]

Conflicting interests are normal. The need for a third party reserves report is driven by its function as the solution to the inherent conflict between the interests of the investor or lender and the management regarding the value of oil and gas assets. Property tax consulting seeks to establish a fair market value in the context of the conflicting interests of the tax payer and the

[4] Martin & Schinzinger: *Ethics in Engineering (4th Ed)*: McGraw Hill Education (India) Private Limited, New Delhi, 2014: p. 159.

taxing entity. Ethical consideration of conflicts of interest seeks to avoid situations in which the professional work product of the evaluator might be unduly influenced or lack independence. Further, it seeks to avoid the appearance of a conflict of interest by disclosing the facts surrounding apparent conflicts of interest to all of the parties involved.

The principal conflict of interest of concern lies in those situations in which the interests of the evaluator conflict with those of the client or employer. Examples include but are not limited to the following:

- An evaluator with a direct or significant ownership interest in the assets being evaluated
- An evaluator who is actively competing to acquire an acreage position in the same area as an employer or client
- An evaluator with a direct or significant interest in an entity that would rely on the evaluation for financing
- An evaluator utilizing proprietary or confidential client/company information for the benefit of the evaluator or other clients

Any situation in which the evaluator stands to gain (or lose) from the decisions and/or actions of the client or employer represents an actual conflict of interest. Generally, these conflicts cannot be adequately addressed by simply disclosing their existence. In these situations, it may be impossible to undertake an evaluation assignment.

Conflicts of interests can also arise between one client and another. The most obvious of these would be a request from two different clients to evaluate the same property. Others might involve a dispute between two or more clients. The general guidelines to addressing these potential conflicts of interest is to completely disclose the nature of the projects to all involved and obtain permission from all involved before proceeding.

Petroleum evaluations are frequently prepared for public regulatory filings. In this case, a conflict of interest would arise if the evaluation was prepared by an employee of the regulatory agency. For example, it would be inappropriate for an employee of a government agency to provide professional services for work product that would subsequently be reviewed by that same agency.

Finally, can a petroleum evaluator issue an independent report for a publicly traded company in which the evaluator owns stock? The answer depends largely upon the nature of the ownership interest. If the ownership interest is sufficient to convey to the evaluator a controlling interest, then a conflict of interest arises that cannot be surmounted by disclosure. Similarly, if the ownership interest is a significant portion of the evaluator's portfolio then that evaluator should decline to prepare the report. For cases in which the evaluator neither owns a controlling interest nor has a significant portion of the evaluator's portfolio invested in the company then disclosure of the existence of the interest is usually sufficient.

In addition to preventing self-dealing for unjust enrichment, conflict of interest principles protect the integrity of the evaluator's independent work product. Where any of the previously mentioned conflicts exist, the independence of the evaluation can be impaired. For a petroleum evaluation to be truly independent, the evaluator's compensation must not be dependent upon the conclusions of the evaluation or the ultimate sales price or appraisal of the property.

Confidential Information

In the course of evaluation of reserves and economics, the professional evaluation engineer is required to deal with and have in possession confidential information of the clients and/or the company of the evaluator. This confidential information includes, but is not limited to, business models, finances, ownership, non-public reservoir and production data, product marketing arrangements, proprietary technology and methods, etc. It is the duty and obligation of the professional reserve evaluator to safeguard this confidential information. Information should not be shared, utilized or marketed for other clients unless expressively approved by the client that controls subject information.

In this era of resource plays, knowledge, technology and methods are being advanced at a significant pace and there is a divide as to what is public sector technology and what is retained as confidential. The professional reserve evaluator must discern the differences and protect the confidential information of his or her client.

Contingent Fee Arrangements

Article IV °5 of the Principles of Acceptable Evaluation Engineering Practice expressly prohibits a member from "seek[ing] or accept[ing] a contingent fee arrangement for preparing a report, giving a professional opinion or providing legal testimony." This section addresses the preparation of a sales offering in which no representation of independence is made by the petroleum evaluator.[5]

Contingent fee arrangements, frequently used for sell-side petroleum evaluations in support of property sales offerings, involve a fee paid to the evaluator that is based either in whole or in part on the total compensation received by the seller. These evaluations are not independent appraisals of the properties. Ethically, contingent fee arrangements should provide the following;

- Disclosure of the fact that the preparer of the evaluation is receiving a contingent fee

[5] This section does not seek to modify in any way Article IV 5 of the Principles of Acceptable Engineering Practice. Rather it clarifies that the evaluation work product prepared under a contingent fee arrangement falls outside of the definition of "report".

- An explicitly stated requirement that the buyer is relying on its own evaluation and not the evaluation prepared pursuant to the contingent fee arrangement
- Reserve classifications used should be based on prevailing practice and regulation
- Production and cost performance should be supported by actual results or sound engineering estimates

It should be clearly and unequivocally stated that no reserves evaluation prepared under a contingent fee arrangement should be relied upon unless it is reviewed and approved by a qualified and totally independent evaluation engineer. Professional opinions and legal testimony may never be sought nor accepted in a contingent fee basis.

Incentive Performance Programs

Incentive performance programs that use reserve-based metrics can be problematic. Incentive performance programs are often used by companies to align employee performance with shareholder or senior management objectives. Unless the company uses an independent third party to prepare its annual reserve report, this may create a situation in which the employees who prepare the company's reserve reports may have a financial interest in the amount of reserves reported. Companies instituting plans that tie incentive payments to reserve booking performance should put management systems in place that address these conflicts and maintain the independence of the final reserve report. This is often accomplished by placing the content of the final reserve report under the control of a team of qualified reserves evaluators who are not incentivized based on the company's reserve bookings.

Conflict of interest, whether actual or apparent, can greatly diminish the integrity of the petroleum evaluator along with the independence of the evaluator's work product. Care must be taken to strictly avoid actual conflicts of interest and to adequately disclose the facts surrounding the appearance of conflicts.

Evaluator Work Product: Control and Proper Use

The lack of proper control of an evaluator's work products can lead to breeches of confidentiality and/or poor financial decisions.

Before the age of the internet, such breeches could be prevented by always keeping tight control on the physical documents and, when finished, exercising proper disposal techniques (e.g., the use of a shredder). This is still true. However, in recent times, there have appeared many examples around the world of breeches of expected conduct by officials with confidential electronic information. Up until those breeches, the victims of such misuses (often the public at-large) had assumed that previous standards of control used for confidential paper information could and would also be adequate for confidential electronic information. This has not been proven to be correct and reserve evaluators must make efforts to safeguard the security of the electronic client information they utilize and the electronically stored and transmitted evaluations they perform.

Also, experienced evaluators can recite stories of their work being used in ways they did not intend. Even when not done for illicit reasons, the decisions made on the basis of work-in-progress or preliminary findings can be the exact opposite of those that would be made by the same decision maker with the final work product in hand. Therefore, it is highly recommended that all transmissions of evaluations or portions of evaluations that are not considered to be "Final" should be accompanied by a qualifying statement that reads something like, "Any information contained herein represents PRELIMINARY work-in-progress unless otherwise noted."

The Petroleum Evaluator and the Environment

The worldwide cultural awareness of the potential effect of any industrial activity on the environment is now a given. It is recognized by the committee that there has historically been very little role for the reserves evaluator in informing the public or an evaluation end-user about environmental risk. In fact, many third party reserve reports that are currently published have specific language that states that environmental issues were not considered at all in the evaluation. The committee agrees that in most circumstances this is appropriate. However, when the reserves being evaluated have a commerciality that is highly dependent upon large-scale technological and/or industrial activity wherein both real and conjectural environmental issues could affect the commerciality of the reserves, as in LNG, deepwater and large-scale unconventional developments, it is the committee's belief that the evaluator should more pro-actively recommend that a professional assessment of the environmental risk associated with the exploration and production activity being contemplated.

The thought behind this is that individuals and/or small entities may from time to time be afforded the opportunity to invest and participate in large scale capital projects such as deep-water drilling. They may make their decision to invest on the basis of a reserves evaluation without realizing that their potential liability in an environmental mishap may not only exceed the value of the project, but may exceed their own entire asset value. In other words, the scale of potential liability of the project dwarfs the potential value of the project and renders the reserve evaluation as potentially misleading unless the evaluator has specifically stated his or her lack of expertise in environmental assessment and made the recommendation that a professional assessment of the environmental risk associated with the activity be performed and considered in conjunction with the reserves evaluation before any major decisions are made on the project.

The Petroleum Evaluator's Duty to Disclose

The reserves evaluator has an ethical duty to disclose any potential conflicts of interest to his or her intended reading audience. Therefore, it is important that any reserves evaluation states very clearly who the intended audience is and for what purpose the evaluation was performed. If left unstated, it will no doubt be open for misuse that could lead to material losses by the public at-large, whether intentional or otherwise.

Just as important as defining the audience and purpose for the evaluation is to then disclose any possible conflicts of interest that the audience should be aware of when considering the evaluation. We can then more pointedly state that the reserves evaluator has the ethical duty to disclose any real or potential conflict among the intended audience of the evaluation.

In addition to conflict of interest disclosures, the reserve evaluator has the ethical duty to ensure that there are no misrepresentations or omissions of important information about the properties that are evaluated. Therefore, it is important that the following information be clearly stated in any reserves evaluation:

- Ownership – Who owns the reserves being evaluated at the time of the report and what is their quantitative ownership including royalties, overrides, promotes, drill-to-earns, etc? Who were the previous owners (if germane)? Is there any relationship whatsoever between the reserves evaluator and the past, present or potential owners of the reserves?
- Effective Date – The report should clearly specify the date from which the analysis is conducted.
- Product Pricing – What is the source and quantitative listing of future product prices and price adjustments used in the evaluation? Were the pricing adjustments calculated by the evaluator from historical financial and accounting data or were they supplied to him or her and, if so, by whom and in what source document and what test did the evaluator make of their reasonableness? If historical data was used to calculate the assumed pricing adjustments, what time period did the data cover and what methodology was used to do the calculation?
- Operating Expenses – What were the operating expense assumptions used in the evaluation? Are there any likely additional operating expenses associated with the reserves that are not included in the evaluation? Were the operating expense assumptions calculated by the evaluator from historical financial and accounting data or were they supplied to him or her and, if so, by whom and in what source document and what test did the evaluator make of their reasonableness? If historical data was used to calculate the assumed operating expenses, what time period did the data cover and what methodology was used to do the calculation?
- Capital Expenditures - What were the capital expenditure assumptions used in the evaluation? Are there any likely additional capital expenditures associated with the reserves that are not included in the evaluation? Were the capital expenditure assumptions calculated by the evaluator from historical financial and accounting data or were they supplied to him or her and, if so, by whom and in what source document and

what test did the evaluator make of their reasonableness? If historical data was used to calculate the assumed capital expenditures, what time period and projects did the data cover and what methodology was used to do the calculation?

- Shrinkage and Plant Products - What were the shrinkage and plant product assumptions used in the evaluation? Are there any likely additional activities or contractual changes that could change the shrinkage or plant products associated with the reserves that are not included in the evaluation? Were the shrinkage and plant product assumptions calculated by the evaluator from historical financial data, plant statements and accounting data or were they supplied to him or her and, if so, by whom and in what source document and what test did the evaluator make of their reasonableness? If historical data was used to calculate the assumed shrinkage and plant products, what time period did the data cover and what methodology was used to do the calculation?

In addition to reporting of conflicts of interests and ensuring that there are no misrepresentations or omissions of important information about the properties that are being evaluated, the reserve evaluator has the ethical duty to make his or her audience aware of any of the following possibilities that he or she may encounter during the course of performing the evaluation:

- Manipulation of the data that is used to evaluate the reserves
- Theft of reserves or reserves revenues or profits
- Unfairness in reporting of reserves or reserves value
- Insider trading by buying or selling the reserves or being an agent to buying or selling the reserves while in possession of material, non-public information about the reserves
- Failure to comply with applicable laws, regulations and guidelines

Ethical Considerations for Expert Witnesses

Editors' Note:

"The Ethical Considerations Involved in Expert Witnessing", written by Scott Hickman, P.E. was first published in the Discussion and Guidance on Ethics prepared for the Society of Petroleum Evaluation Engineers, *May 2005. Mr. Hickman reviewed this manuscript in 2017, and has authorized its continued use. Mr. Hickman writes from his considerable experience as a petroleum engineer, evaluator, and expert witness. This article reflects his experience from his United States based engineering practice, but we believe that the ethical principles that underlie this discussion are applicable across a broad range of jurisdictions.*

THE ETHICAL CONSIDERATIONS INVOLVED IN EXPERT WITNESSING

The *Code of Ethics of Engineers* and the *Principles of Acceptable Evaluation Engineering Practice* found in Appendices A and B, respectively, of the SPEE By-Laws set a reasonable and comprehensive standard of conduct for engineers engaging in the various aspect of reserve evaluation. Are any other standards needed for an engineer serving as an expert witness in civil litigation? In theory no, but in practice yes. Both the logic and procedures involved in civil legal proceedings are foreign enough to the inexperienced engineer witness to create numerous pitfalls even for the most conscientious person. Not only can an engineer unknowingly violate codes of conduct, but could be guilty of contributing to a miscarriage of justice. The connection between an evaluation engineer's normal activities and the public's welfare may often seem tenuous, but not when it involves the justice system. What greater indictment of an engineer's professionalism can there be than impeding the administration of fair, impartial justice, which is the very foundation of a free society.

This paper's focus is on the ethics of expert witnessing, not how to be an effective witness. However, there is considerable overlap between the two topics. The expert is being neither effective nor ethical if not fully prepared within the time and job scope constraints of the assignment. Understanding how the civil justice system works in America is crucial to being both an ethical and effective witness. In the simplest terms, the justice system allows each party in the dispute to bring forth witnesses and the introduction of evidence that show all the facts and expert opinions that will help make their case. Each party is also allowed to examine and question all the evidence and witnesses from the other side. After all the facts and opinions have been presented and examined in open court, the impartial trier-of-fact (whether judge or jury), after careful deliberation, will decide which party has the "preponderance of evidence" in their favor. Thus justice is served in an American court.

This adversarial approach to civil litigation creates a confrontational and at times hostile atmosphere that causes most of the pitfalls for expert witnesses. It is extremely important to understand at that role of each character in the drama of a legal proceeding. The judge maintains order in the court, rules on points of law and procedure that arise and instructs the jury on what issues of fact that they are to decide. In a "bench" trial the judge also acts as the jury. Each party in the case will have one or more attorneys representing them. While attorneys are officers of the court and held to certain standards of conduct they are also hired to be advocates for their client's case. It is their job to bring forth every possible fact and opinion that could further their client's cause while questioning the validity of facts and opinions and the credibility of witnesses presented by the other side. **Attorneys are not impartial.**

In contrast, an expert witness brings some specialized skill, knowledge, experience, education, and training into the courtroom or hearing room to assist the trier-of-fact. The expert should be impartial, rendering opinions based on the facts. The expert is also human so being completely impartial and wholly independent are somewhat idealized concepts in the confrontational and often emotionally charged atmosphere of a trial. It often takes a well-developed sense of professionalism to maintain objectivity and function as an ethical witness in these circumstances. The special circumstances of civil litigation raises issues that the engineer does not normally encounter, hence the justification for a presentation devoted to the ethics of expert witnessing.

Assuming the engineer has at least a general understanding of the civil legal system procedures, the ethical considerations start with the decision on whether to accept the engagement as an expert witness. Are you qualified by training and experience to evaluate and opine on the technical issues that are involved? Do you have any conflicts of interest with parties in the case? There are direct conflicts such as 1) having worked for the opposing party on the issues involved or 2) you are on record as stating an opinion that may conflict with opinions you may reach in this case and your participation in the case could place that party at a disadvantage. There can be potential conflicts that arise out of established relationships that may require a business or personal rather than ethical decision. As with any consulting assignments where independent opinions are given, compensation should not be based on the outcome of the lawsuit.

There are other less tangible issues to be considered before accepting the assignment. Are you comfortable with the reputation of the party you would be representing? An expert can be a completely ethical witness even working for someone of questionable repute. In America everyone has the right to hire the best available legal counsel and technical assistance for their "day in court." But if you lie with dogs you're liable to get fleas (advice given me by a disbarred lawyer). There can be undue pressure to slant your opinions and less than full disclosure of all the facts and data. At best it is an uncomfortable situation and can become a quagmire, particularly for the inexperienced witness.

When considering a litigation assignment the issues involved in the lawsuit are a legitimate concern. This is a gray area. It is the court's, not the expert's, responsibility to pass judgment on the issues. However, if you are uncomfortable with the prospective client's legal or factual position to the extent that it could impact your performance or compromise your principles then it does become an ethical decision. Similarly the job scope assigned to the expert can have ethical ramifications. Normally the expert is not given carte blanche to study all the issues involved that pertain to his area of expertise. Logically the client and its legal counsel will request that the expert investigate and opine in certain areas only. If the charge is so narrowly focused that it could distort the issues, it's the expert's professional responsibility to discuss his concern with counsel then render a decision as to accepting or continuing the assignment.

On the other side of the coin, if during the course of his investigation the expert develops information or forms opinions that could be detrimental to the client's position; he has the obligation to **verbally** inform counsel. Depending on the circumstances, the decision may then be made not to use the expert's services. If so, the expert has the duty to maintain the usual client confidentiality, even if he is no longer involved in the case. Often a prospective expert witness is requested to do a preliminary review up front on the important issues to see if he thinks the facts will lead to conclusions and opinions supportive of the client's position. Keep in mind that once such a review has been done you are ethically obligated to maintain confidentiality even if you do not get the assignment and cannot work for other parties in the lawsuit.

The law in Texas classifies experts either as consulting or testifying experts. This allows the client and attorney to freely consult with an expert and obtain his or her true opinion. If not favorable, they may drop the case, modify their position or change experts. The work product of a consulting expert is protected from disclosure. On the other hand, all of a testifying expert's work product and most of the written communications with the client and legal counsel are subject to discovery by other parties in the lawsuit. Often an engineering expert serves in both capacities; consequently all of his work product is discoverable. It is illegal to not disclose or to destroy information that has been requested through discovery. You may also be cross examined under oath during deposition or trial about what you have investigated and concluded in the course of your assignment. It is very important not to write reports, conclusions or opinions without specific instructions from the client's legal counsel.

Expert witnesses get into questionable ethical positions unknowingly by not clearly understanding their role in relation to the attorneys' role. Remember that attorneys are advocates for their clients. Short of knowingly putting on false testimony attorneys are largely free to explore every alleged fact, conceivable theory or half-baked opinion that would support their client's position, while questioning the credibility of every aspect of the opposition's case. Intuitively an expert views the opposing attorney as the enemy who will question his competency and opinions through trick questions. In truth, if a qualified expert has done his homework and is not trying to withhold facts or play mind games and answers **only** the question asked, the opposing attorney is no threat.

The **main danger** an expert faces is from his client's attorney who may pressure you about your opinions or suggest revisions in your testimony to be more "responsive". You may be asked to stretch your expertise into areas where you aren't fully qualified. There is nothing necessarily illegal or unethical about attorneys doing this. They are fulfilling their advocate role, but the expert as an unbiased, independent party has the professional responsibility to decide what subjects he can opine on and to state his opinions clearly and fully. If you are not able to withstand the power of suggestion from an aggressive attorney, it would be wise not to serve as an expert witness.

While in theory an expert is an unbiased, independent party, it is human nature to invest in your own credibility and to want your side to win with you contributing to their success. After all you are part of the "team". This is particularly true in complex cases where you have spent many long hours in the presence of clients and attorneys, serving both as consultant and testifying expert. The team spirit can really thrive under these conditions, but an ethical expert cannot let this impair his professional judgment even if it strains relationships. You are not really a member of a support team in the sense that a purely consulting expert who is not testifying would be. While a consulting engineer is always held to a professional code of conduct, he can qualify his opinions in a report by disclaimers, disclosures and limited usage clauses. The lack of such a safe harbor places a special burden on a testifying expert witness to evaluate information, interpret facts and render opinions in an impartial manner that will help the court understand technical issues. **The dual role of consulting and testifying expert along with aggressive advocacy by the client's lawyer create the major pitfalls for a would-be ethical witness.**

In the area of reservoir engineering and reserve valuation, the subject matter often tends to be more gray than black and white, requiring varying degrees of subjective judgments by the practitioner. How does the expert reconcile this subjectivity with the duty to help the court understand technical issues? Opposing attorneys love to play to the jury by expressing "shock" at an expert's admission that reserve volumes are estimates rather than exact measurements. This response conveniently overlooks the fact that industry uses such estimates to conduct its normal business. It's up to the expert to convey this to the court without over or understating the accuracy involved.

Subjectivity also requires that the expert stick to procedures generally accepted by industry as opposed to utilizing some nonstandard approach for the occasion of litigation. The growth of the contingency fee litigation industry in the 1980's gave rise to hundreds of lawsuits claiming certain products caused harm to the plaintiffs. Often these claims were based on little more than junk science and testimony by a "for hire" expert witness. This resulted in a number of court decisions about the acceptability of expert scientific testimony; most notable of which was the Daubert decision upheld by the U.S. Ninth Circuit court in an opinion issued January 1995. Several states have adopted the Daubert approach, which can be illustrated by the Supreme Court of Texas' reasoning for excluding an expert's testimony:

1. was not grounded upon careful scientific methods and procedures,
2. was not shown to be derived by scientific methods or supported by appropriate validation,
3. was not shown to have a reliable basis in the knowledge and experience of the witness' scientific discipline,
4. was not based on theories and techniques that had been subjected to peer review and publication,
5. was not based on a procedure reasonably relied upon by experts in the field.

A more detailed treatment of the ramification of the Daubert decision can be found in an article written by Harvey G. Brown, Jr. for the 25th Annual Advanced Civil Trial Course, entitled "UPDATE ON *DAUBERT* CHALLENGES EXPERTS". Brown elaborates on the eight gates that must be passed to make an expert's testimony admissible in court.

The subdiscipline of reservoir engineering is based on valid scientific principles and industry accepted practices that are the subject of continuous peer review and publication. It requires more subjective judgments than most other engineering disciplines due to the lack of sampling from the object, i.e. the reservoir being analyzed. The difficulty this presents depends on the quantity and quality of the information in relationship to the heterogeneity of the reservoir. Daily, engineers successfully analyze reservoirs and evaluate reserves using training, experience, and sound judgment. Expert witnesses unable to apply these attributes competently tend to demonstrate an *"the exact answer isn't known so my opinion is as good as your opinion"* attitude which does not help the court to understand technical issues. Certainly two competent, unbiased engineers can look at the same set of facts and derive different conclusions, but they should also be able to verbalize where and why the differences occur. This gives the court a basis for making a decision rather than being faced with two intractable opinions. There are many engineers evaluating reserves that have never advanced beyond the "cookbook" approach. They may be able to function quite capably within restricted conditions, **but they do not belong in court as expert witnesses.**

Since the civil justice system is based on the opposing parties confronting each other in open court after a discovery process, the experts have the opportunity to "peer review" each other's opinions both during deposition and in court room testimony. Weaknesses or apparent inconsistencies in the testimony can be pursued by direct testimony and particularly during cross examination. On occasions both sides may not be equally represented by capable experts, which can remove the "peer review" constraint for one side. Obviously, it is not the opposing expert's duty to make the case for the other side. Each party is responsible for being adequately represented. However it would be unethical to take advantage of the lack of "peer review" in such a situation by offering questionable or unintelligible testimony.

The number and nature of ethical considerations involved in expert witness testimony may be surprising to some. Yet all the issues discussed arise directly out of the responsibility that a professional engineer has toward protecting the public health, safety, and welfare. For an engineer to accept an expert witnessing assignment without being fully aware of the ethical issues and responsibilities involved is both unprofessional and unethical.

In summary, **it is the expert's responsibility to help the court understand technical issues**. However the expert offers his testimony in response to questions asked initially by his attorney (direct) followed by questions from the opposing attorney (cross). Expert testimony is not an opportunity **for extemporaneous** speaking.

Cases

Competence

Situation

A consulting petroleum engineer receives an inquiry from a prospective client about performing a resource/reserves evaluation for properties in a shale play. Although the engineer has spent many years as a petroleum evaluator, the evaluator's current business is focused primarily on general reservoir engineering consulting. The unconventional boom occurred after the engineer switched focus, so he has not evaluated any unconventional plays.

The engineer has never done work for this prospective client, but they know each other as former co-workers. The prospective client is aware of the engineer's lack of current unconventional play evaluation skills, but is still interested in using the engineer for the work.

What are the possible courses of action for the engineer? How will the engineer decide which action to take?

Analysis

This case involves the competence of the consulting engineer. The issue is that the engineer may not possess the current skills relevant to the task required by the prospective client, which in this case involves evaluating oil and gas properties containing unconventional resources. Mitigating conditions: (i) the engineer has many years of experience valuing petroleum properties, (ii) the potential client is aware of the engineer's deficiencies.

One of the Fundamental Canons of Engineering (reference) states that engineers shall "perform services only in areas of their competence."[6] Requirements for professional engineers in many jurisdictions have some language about competency. As an example, the Texas Board of Professional Engineers gives the guidance that, "the licensed engineer is bound to only practice engineering in areas where competent, trained, and qualified or may be subject to enforcement actions."[7] Clearly, the expected conduct of professional engineers is that they will only perform work for which they are competent. This raises the question of how a practicing engineer can gain a new competence.

The challenges of industry are constantly changing and so the technology and techniques to deal with those challenges are also constantly changing and improving. Consequently, even the best engineers must regularly update their skills in order to stay current and competent. The existence of numerous industry seminars, courses, and forums attest to this reality. A key point to recognize is that these resources are designed to provide "incremental" training to an engineer who is already competent in some areas. In other words, a solid foundation of knowledge and experience is necessary in order to gain benefit from the training. This means, for example, that an engineer who has performed conventional oil and gas evaluations in the past could reasonably be expected to be able to upgrade his skills "on the job" or from readily available self-learning sources to perform an evaluation of unconventional resources. But that same engineer couldn't reasonably be expected to be able to design an offshore production facility without significant additional formal training and experience.

The engineer has two choices. One can decline the work, referencing a lack of proficiency at the task, or accept the work with the caveat that one will be learning the task as the evaluation is performed.

The following actions are recommended: 1) The engineer should first honestly assess his skills related to the task. If he recognizes deficiencies, but believes that his skill gap can be closed by simply updating or extending current skills, it is reasonable to expect that the engineer would be able to perform the task competently. 2) The engineer must make his specific competency level clear to the prospective client, allowing the client to decide whether to proceed. 3) If the engagement proceeds, and if the engineer is paid on a time based fee, the engineer should not charge the client for time spent specifically on updating his skills.

---

[6] https://www.nspe.org/resources/ethics/code-ethics

[7] http://engineers.texas.gov/lic_faq.htm

Control of Use of Work Product

Situation

A petroleum evaluator receives a letter from a bank regarding some earlier work done for a client. The bank was in the final stages of authorizing a loan and they were unable to locate the cover letter for the reserve report they had in their file. The bank requested that the evaluator send either a copy of the cover letter or a replacement reserve report for the bank's file.

The bank's request presented two problems for the evaluator. The person applying for the loan was not the evaluator's client, and the "reserves report" was in fact an upside sensitivity case that had been prepared at the request of another client.

What should the evaluator do? Which parties should the evaluator notify? Should the evaluator have prepared the upside case, since it did not reflect the evaluator's estimate of proved reserves?

Analysis

This case involves the proper use of the reserve report. From the facts, we can identify four issues: (i) the evaluator provided the client with an upside case that portrayed and overstated value for the properties analyzed, (ii) the client allowed a copy of the upside sensitivity case to leave their control, (iii) a third party who was not a client of the evaluator attempted to use this upside case to secure financing without the knowledge or consent of the evaluator, and (iv) the bank began their process without an actual report in hand.

Analyses that do not comply with reserve reporting guidelines may be prepared to better understand the range of outcomes - both high and low - that may occur in a petroleum project. While there is no ethical barrier to preparing such cases, it is incumbent upon the evaluator to conspicuously disclose that this work product does not conform to reporting guidelines and should not be used for the purposes typically associated with reserve reports.

Clients are ethically bound to take care that petroleum evaluation work prepared for them is used solely by them, their finance sources, and their agents for the purposes stated in the report. SPEE Recommended Evaluation Practice #1 - Elements of a Reserve and Resource Report §2.1 states, "The preparer should clearly state the purpose for which the report is being prepared and for whom it is prepared[8]." These limitations should be honored by the client.

In addition to being discourteous, the use of petroleum evaluation reports that were prepared for others without their consent carries with it serious ethical and practical issues. Leaving aside the obvious issue of the use of another person's intellectual property without compensation, reserve reports are written for a specific time and for specific clients. Conditions may have materially changed, and the financial capability of the owners is a factor in reserve classification. None of these are honored when using a reserve report in this manner.

The bank's processes were ultimately successful. The issues were identified and addressed. The process might have been more efficient had the bank required that the prospective borrower supply a current reserve report prepared by a competent evaluator for the prospective borrower before continuing to process the loan.

The evaluator should do the following: (i) Notify the bank that the document they have in hand is not a reserve report and that it was not intended to be used for that purpose. (ii) Notify the client that work product prepared for their use has been presented by a third party in an attempt to secure a loan. (iii) If the client approves, notify the third party that the document they presented has been specifically disclaimed.

---

[8]Society of Petroleum Evaluation Engineers: Recommended Evaluation Practice #1: Elements of a Reserve and Resource Report (Version 1.1): Society of Petroleum Evaluation Engineers, Houston, Texas, 2012. https://secure.spee.org/sites/spee.org/files/rep01_-_reserve_report_contents_v1.1_final.pdf

Risk to the Public (Unsophisticated Investor)

Situation

A land developer hires a small third party reservoir engineering firm to prepare a "Reserves Report" of his 30,000 acres of contiguous oil & gas leases located in an, as yet, undrilled fringe area of an unconventional play. The land developer intends to use the report to attract investors to drill.

An evaluation engineer working for the third party reservoir engineering firm is handed the project and begins the analysis by gathering data from public and subscription services. The evaluation engineer finds there are numerous old vertical wells with publicly available logs on the acreage that confirm the presence of the target shale. In addition, there are old scout tickets on many of these vertical wells that give many inferences of hydrocarbons encountered during the drilling of the wells. However, there is no historical production from the interval anywhere within 10 miles of the leases.

The evaluation engineer compares the thickness and log characteristics of the target shale across the land developer's acreage to the thickness and log characteristics of the target shale in the closest development, some 10 miles away, and finds them to be similar. The evaluation engineer composes an evaluation of the land developer's acreage on the basis of the distant horizontal well development by a major independent and classifies all of the hypothetical hydrocarbon volumes in it as "Contingent Resources". Also included is the Petroleum Resources Management Systems (PRMS)[9] definitions of all categories of Reserves and Resources in the report and makes it clear in its narrative why one cannot assume any of the hypothetical hydrocarbon volumes he has estimated can be considered as "Reserves".

The land developer gives a copy of the evaluation to a group of wealthy individuals who play golf together regularly and go to the same church but have no knowledge of the upstream petroleum industry. The pastor of the church they attend is also always invited to play golf with these individuals at no expense to him, an opportunity he partakes in regularly. The golfing buddies all discuss the opportunity on one of their outings and decide to fund the Land Developer's plans. The pastor, upon hearing the others discuss the opportunity, asks for a copy of the third party evaluation and, eventually asks the wealthy golfing buddies if it would be possible to participate with a comparatively modest investment. The pastor is allowed to participate and contributes almost all of his savings.

The land developer obtains funding from the golfing buddies and the pastor to acquire 3D seismic and drill several horizontal wells. The 3D is interpreted by a contract geophysicist.

---

[9] Society of Petroleum Engineers, American Association of Petroleum Geologists, World Petroleum Council, Society of Petroleum Evaluation Engineers: Petroleum Resources Management System: http://www.spe.org/industry/docs/Petroleum_Resources_Management_System_2007.pdf: Accessed May 9, 2017.

Another contract geoscientist is hired to work up the well proposals. The geologist prepares and submits several well proposals to the golfing buddies, the pastor and the land developer. All of the wells are drilled and completed using contract operations personnel. None of the wells produces even 10% of the volumes stated as Contingent Resources in the third party evaluation.

The golfing buddies, the pastor and the land developer ask the evaluation engineer to present to them "what went wrong." The evaluation engineer, who has had no knowledge of the properties since the day he completed his evaluation, is given the well data and structure maps on several seismic events above and below the unconventional resource interval. The structure maps show the horizons below the target interval to be "swarming" with faults that are mostly not present on the maps of the horizons above the target shale.

What should the evaluator do?

Analysis

The evaluator should present his original evaluation and thoroughly explain the PRMS definitions that were included in the original evaluation. If desired, the evaluator can then give possible reasons for the commercial failures of the wells drilled to date and make a personal recommendation for further action on the project.

Although in this case the evaluator fulfilled his responsibility and did what is required to ensure the evaluation was not misunderstood, the project still caused the financial ruin of an unsophisticated investor. Perhaps, the evaluator could have been clearer about the purpose and use of the original evaluation in the text or made a very clear and straightforward statement about the risk of the project. However, the upstream oil and gas world is extremely risky and, no matter how good a job we do as evaluators, commercial failures will occur.

Code of Ethics of Engineers[10]

The Fundamental Principles

Engineers uphold and advance integrity, honor and dignity of the engineering profession by:

1. using their knowledge and skill for the enhancement of human welfare;
2. being honest and impartial, and serving with fidelity the public, their employers and clients;
3. striving to increase the competence and prestige of the engineering profession; and
4. supporting the professional and technical societies of their disciplines.

The Fundamental Canons

1. Engineers shall hold paramount the safety, health and welfare of the public in the performance of their professional duties.
2. Engineers shall perform services only in areas of their competence.
3. Engineers shall issue public statements only in an objective and truthful manner.
4. Engineers shall act in professional matters for each employer or client as faithful agents or trustees, and shall avoid conflicts of interest.
5. Engineers shall build their professional reputation on the merit of their services and shall not compete unfairly with others.
6. Engineers shall act in such a manner as to uphold and enhance the honor, integrity and dignity of the profession.
7. Engineers shall continue their professional development throughout their careers and shall provide opportunities for the professional development to those engineers under their supervision.

---

[10] Source: https://secure.spee.org/joining-spee section related to "Important SPEE Documents", originally published by the National Society of Professional Engineers

About the Society of Petroleum Evaluation Engineers

The Society of Petroleum Evaluation Engineers (SPEE) is an international professional society dedicated to advancing the art and science of quantifying, classifying, and valuing oil and gas reserves and resources.

Mission Statement

Recognizing that Petroleum Evaluation Engineering is a specialized field, the Society is dedicated to the promotion of professional growth of the membership and to the advancement of the profession of Petroleum Evaluation Engineering by demonstrating by example the highest standard of ethics, by promoting continuing education of our membership and by education of the public in the area of oil and gas reserve definitions, reserve evaluations, and fair market value.

About the Society

Founded in 1962 by petroleum engineers Harold Vance, William Hurst, and H. F. Poyner, Jr., the Society of Petroleum Evaluation Engineers was established to bring together practitioners in this highly specialized area of petroleum engineering. The professional activities of SPEE members are guided by by-laws that require the highest ethical standards that address the relationships of members with the public, with employers, with clients, with other members, and with SPEE.

In cooperation with other industry groups, SPEE has worked to standardize oil and gas reserve definitions and provide application guidelines for a variety of settings. The SPEE Monograph Series (I-IV) and Recommended Evaluation Practices Series (1-10) address reserves definitions, fair market value, unconventional resources, and various technical aspects of reserve reporting and valuation. The Calgary Chapter of SPEE holds the copyright for the Canadian Oil and Gas Evaluation Handbook, a key reference for oil and gas reserve reporting in Canada. Since 1982, SPEE has published an annual comprehensive survey of the key economic parameters used by petroleum evaluators.

Continuing professional development is provided by the Society's Annual Meeting, as well as through more frequent meetings of the eleven local Chapters operating in the United States, Canada, and Europe.

For more information, please visit SPEE.org.

Resources

Select Professional Societies Ethics Resources

Society of Petroleum Evaluation Engineers

Code of Ethics of Engineers

Principals of Acceptable Evaluation Engineering Practices

Discussion and Guidance of Ethics (May, 2005)
Http://www.spee.org/resources/ethics-paper

Society of Petroleum Engineers

Code of Conduct: http://www.spe.org/about/professional-code-of-conduct.php

National Society of Professional Engineers

http://www.nspe.org/resources/ethics

National Institute for Engineering Ethics

www.niee.org

Royal Academy of Engineering

http://www.raeng.org.uk/policy/engineering-ethics/ethics

Petroleum Reserves Evaluation Resources

Society of Petroleum Evaluation Engineers

https://secure.spee.org/

Reserve Definition Committee (Up to date information is available on the web site. (World, Canada and USA))

Recommended Evaluation Practices (REPs)

Monographs

1.) "Guidelines for Application of the Definitions for Oil and Gas Reserves."

3.) Guidelines for the Practical Evaluation of Undeveloped Reserves in Resource Plays."

4.) "Estimating Ultimate Recovery of Developed Wells in Low-Permeability Reservoirs."

Society of Petroleum Engineers

http://www.spe.org/industry/reserves.php

Petroleum Resources Management System

Guidelines for the Application of the Petroleum Resources Management System

Laws and Regulations Governing Petroleum Reserve Evaluation

The Society of Petroleum Evaluation Engineers strives to maintain up to date information including:

Canadian Regulations

Canadian Oil and Gas Handbook (Developed in part through the efforts of the Calgary Chapter of the SPEE.)

United States

Security and Exchange Commission

Internal Revenue Service of the Treasury Department

Financial Accounting Standards Board

States Oil and Gas Tax Rates

Made in the USA
Monee, IL
30 April 2026

49234287R00026